FLORA POETICA

FLORA POETICA

PENGUIN BOOKS

The beauty of the Australian bush is staggering. I have been much struck with the subdued and harmonious colours of the landscape … Wherever you look, there are romantic tapestries, and human invention cannot improve that design.

George Lambert

MY COUNTRY

I love a sunburnt country,
A land of sweeping plains,
Of ragged mountain ranges,
Of droughts and flooding rains.
I love her far horizons,
I love her jewel-sea,
Her beauty and her terror –
The wide brown land for me!

Dorothea Mackellar

Archirhodomyrtus beckleri
rose myrtle

THE SONG OF THE BUSH

By the side of a dried-up creek,
In the glare of a scorching sun,
A squatter sat squat upon a log,
Surveying his parched-up run,
Squat – squat – squat;
'T would have made our rulers blush.
To have heard the dolorous words from his lips,
(Oh that their tone would reach Governer Gipps!)
As he sang the 'Song of the Bush.'

'H.B.'

Boronia megastigma
brown boronia

CLANCY OF THE OVERFLOW

In my wild erratic fancy visions come to me of Clancy
   Gone a-droving 'down the Cooper' where the Western
     drovers go;
As the stock are slowly stringing, Clancy rides behind
     them singing,
   For the drover's life has pleasures that the townsfolk
     never know.

And the bush hath friends to meet him, and their kindly
     voices greet him
   In the murmur of the breezes and the river on its bars,
And he sees the vision splendid of the sunlit plains extended,
   And at night the wond'rous glory of the everlasting stars.

A.B. (Banjo) Paterson

Parachidendron pruinosum
raintree

SPRING SONG

I am the Vision and the Dream
   Of trembling Age, and yearning Youth;
I am the Sorceress Supreme.
   I am Illusion; I am Truth.

There is no music like to mine;
   I sing in green, and gold and red;
I pour from secret casks the wine
   That cheers the cold hearys of the dead.

My names are all the names impearled
   In all the songs my singers sing;
I am the sweetheart of the world –
   I am Carissima – the Spring!

Victor Daley

Oxylobium atropurpureum
mountain pea

IN DEFENCE OF THE BUSH

But the bush hath moods and changes,
   as the seasons rise and fall,
And the men who know the bush-land
   — they are loyal through it all.

From *The Bulletin*, 23 July 1892

Dendrobium bigibbum
Cooktown orchid

There are no lines in nature.

(Duncan) Max Meldrum

Darwinia citriodora
lemon scented myrtle

The mystery of the bush seems to recede from you as you advance, and then it is behind you if you look around.

David Herbert
From *Kangaroo*

Grevillea rosmarinifolia
rosemary grevillea

PASSION FLOWER

Choose who will the wiser part –
I have held her heart to heart;

And have felt her heart-strings stirred,
And her soul's still singing heard

For one golden-haloed hour
Of Love's life the passion-flower.

So the world may roll or rest –
I have tasted of its best;

And shall laugh while I have breath
At thy dart and thee, O Death!

Victor Daley

Thelymitra aristata
scented sun-orchid

THE MAN FROM SNOWY RIVER

And down by Kosciusko, where the pine-clad
 ridges raise
Their torn and rugged battlements on high,
Where the air is clear as crystal, and the white
 stars fairly blaze
At midnight in the cold and frosty sky,
And where around The Overflow the reed beds
 sweep and sway
To the breezes, and the rolling plains are wide,
The Man from Snowy River is a household
 word today,
And the stockmen tell the story of his ride.

A.B. (Banjo) Paterson

Actinotus helianthi
flannel flower

A SPRING SONG

The young green leaves is shootin' on the trees,
The air is like a long, cool swig o' beer,
The bonzer smell o' flow'rs is on the breeze,
An' 'ere's me, 'ere,
Jist moochin' round like some pore, barmy coot,
Of 'ope, and joy, an' forchin destichoot.

C. J. Dennis

Kennedia prostrata
running postman

KU-RING-GAI ROCK CARVINGS

Making love for ten thousand years on a rockledge:
The boronia springs up purple
From the stone and we lay together briefly
For as long as those two lovers.

David Gordon Campbell

Acacia leprosa 'scarlet blaze'
red wattle

IDENTIFICATIONS

And sometimes lying there under the night,
wet by a passing shower, we see emerge
from milky cloud, a blown moon in full flight,
then suddenly the trees on either edge,

Stringy and scribbly and apple box
burn with the voltage of a million lives
as though the stars had flown them in flocks
singing and sighing, glittering in the leaves.

Frank Kellaway

Eucalyptus rhodantha
rose gum

HAPPY DAYS

A fringe of rushes, – one green line
Upon a faded plain, –
A silver streak of water-shine, –
Above, tree-watchers twain.
It was our resting-place awhile,
And still, with backward gaze,
We say: 'Tis many a weary mile, –
But there were happy days.

And shall no ripple break the sand
Upon our farther way? –
Or reedy ranks all knee-deep stand? –
Or leafy tree-tops sway? –
The gold of dawn is surely met
In sunset's lavish blaze;
And, – in horizons hidden yet, –
There shall be happy days.

Mary Hannay Foott

Acacia aphylla
leafless rock wattle

THE BUSH, MY LOVER

The camp-fire gleams resistance
To every twinkling star;
The horse-bells in the distance
Are jangling faint and far;
Through gum-boughs lorn and lonely
The passing breezes sigh;
In all the world are only
My star-crowned Love and I.

Will Ogilvie

Brachychiton discolour
lace bark tree

A MIDSUMMER NOON IN THE AUSTRALIAN FOREST

Not a sound disturbs the air,
There is quiet everywhere;
Over plains and over woods
What a mighty stillness broods!

Only there's a drowsy humming
From yon warm lagoon slow coming:
Tis the dragon-hornet – see!
All bedaubed resplendently.

O 'tis easeful here to lie
Hidden from noon's scorching eye,
In the grassy cool recess
Musing thus of quietness.

Charles Harpur

Eucalyptus haemastoma
scribbly gum

BELL-BIRDS

By channels of coolness the echoes are calling,
And down the dim gorges I hear the creek falling:
It lives in the mountain where moss and the sedges
Touch with their beauty the banks and the ledges.
Through breaks of the cedar and sycamore bowers
Struggles the light that is love to the flowers;
And, softer than slumber, and sweeter than singing,
The notes of the bell-birds are running and ringing.

Henry Kendall

Dietes robinsoniana
wedding lily

There are many possible approaches to Australian garden design, and they all reflect the designer's individual response to gardens . . . I like the whole thing to be as wild as possible, so that you have to fight your way through in places.

Edna Walling

Neptunia gracilis
sensitive plant

BEAUTY IMPOSES

Beauty imposes reverence in the Spring,
Grave as the urge within the honeybuds,
It wounds us as we sing.

Beauty is joy that stays not overlong.
Clad in the magic of sincerities,
It rides up in a song.

Beauty imposes chastenings on the heart,
Grave as the birds in last solemnities
Assembling to depart.

John Shaw Neilson

Crowea saligna
pink star flower

SPRING BREEZES

Spring breezes over the blue,
now lightly frolicking in some tropic bay,
go forth to meet her way,
for here the spell hath won and dream is true.

O happy wind, thou that in her warm hair
   mayst rest and play!
could I but breathe all longing into thee,
so were thy viewless wing
as flame or thought, hastening her shining way.

Christopher Brennan

Prostanthera ovalifolia
oval-leaved mint-bush

The hot wind, born amid the burning sand of the interior of the vast Australian continent, sweeps over the scorched and cracking plains, to lick up their streams and wither herbage in its path, until it meets the waters of the great south bay; but in its passage across the straits it is reft of its fire, and sinks, exhausted with its journey, at the feet of the terraced slopes of Launceston.

Marcus Clarke

Crinum flaccidum
Darling lily

WARATAH AND WATTLE

Though poor and in trouble I wander alone,
With a rebel cockade in my hat;
Though friends may desert me, and kindred disown,
My country will never do that!
You may sing of the Shamrock, the Thistle, and Rose,
Or the three in a bunch if you will;
But I know of a country that gathered all those,
And I love the great land where the Waratah grows,
And the Wattle bough blooms on the hill.

Henry Lawson

Telopea speciosissima
New South Wales waratah

THE WATTLE

I saw it in the days gone by,
When the dead girl lay at rest,
And the wattle and the native rose
We placed upon her breast.

I saw it in the long ago
(And I've seen strong men die),
And who, to wear the wattle,
Hath better right than I?

I've fought it through the world since then,
And seen the best and worst,
But always in the lands of men
I held Australia first.

Henry Lawson

Acacia pravissima 'kuranga cascade'
Ovens wattle

Not very far away a number of Banksia men were sitting in a Banksia-tree, basking in the sun and planning mischief.

'Bunch and scrunch 'im!' shouted one.
'Hit and spit 'im!' growled another.
'String and ring 'im!' snarled a third.

Then they all jumped about, grunting and chattering and shaking the bough till the leaves rattled.

May Gibbs
From *The Adventures of Snugglepot and Cuddlepie*

Banksia aemula
Wallum banksia

CAMPHOR LAUREL

Here in the slack of night
the tree breathes honey and moonlight.
Here in the blackened yard
smoke and time and use have marred,
leaning from that fantan gloom
the bent tree is heavy in bloom.

Judith Wright

Hakea macraeana
Macrae's hakea

This country possesses numerous advantages . . .
We enjoy here one of the finest climates in the world.
The necessaries of life are abundant, and a fruitful soil affords
us many luxuries. Nothing induces me to wish for change . . .

Elizabeth Macarthur

Leptospermum squarrosum
peach-flowered tea tree

Being lost in Australia gives you
a lovely feeling of security.

Bruce Chatwin

Phaius tankervilleae
greater swamp orchid

Here tower bright green soft woods smothered in all manner of living parasites, orchids and ferns; looped and twisted with hundreds of feet of great vines, thick as the upper arm.
The jungle teems with exotic birds who never cross the dark threshold into the sunny warmth of the eucalyptus country. Here, too, are colourful outlandish flowers which bloom only in the green twilight.

Bernard O'Reilly
From *Green Mountains*

Banksia coccinea
scarlet banksia

OUTBACK

All the unhallowed beauty I have found;
All free — discordant shrills
and form-defying wonders above ground,
like writhen trees with draggled foliage
struggling along the courses of wayback creeks;
scarlet — and — green
sky — streaking parrot — fires with parrot shrieks
echo — shattering the shoulders of the hills;
and desert — sunset — rage
Rage for my mind, be clamant, do not cease
you are my holiest habitat of peace.

Rex Ingamells

Swainsona formosa
Sturt's desert pea

Every country has its own landscape which deposits itself in layers on the consciousness of its citizens . . .

Murray Bail
From *Eucalyptus*

Scaevola 'new blue'

JOURNEY: THE NORTH COAST

Down these slopes move, as a nude descends a staircase,
the slender white gum trees,
and then the countryside bursts open on the sea –
across its calico beach, unfurling;
strewn with flakes of light.

Robert Gray

Caladenia saccharata
sugar orchid

Earth is here so kind, that just tickle her with a hoe and she laughs with a harvest.

Douglas Jerrold

Caladenia flava
cowslip orchid

GUM-TREES STRIPPING

Wisdom can see the red, the rose,
the stained and sculptured curve of grey,
the charcoal scars of fire, and see
around that living tower of tree
the hermit tatters of old bark
split down and strip to end the season;
and can be quiet and not look
for reasons past the edge of reason.

Judith Wright

Nelumbo nucifera
lotus flower

To the native-born Australian the Wattle stands for home, country, kindred, sunshine, and love – every instinct that the heart most deeply enshrines . . . Let Wattle henceforth be a sacred charge to every Australian . . . Let us rouse our young people's sense of chivalry, and make the Wattle synonymous with Australia's honour.

A. L. Storrie

Acacia penninervis
hickory wattle

DAWN AND SUNRISE IN THE SNOWY MOUNTAINS

A few tin strips of fleecy cloud lies long
And motionless above the eastern steeps,
Like shreds of silver lace: till suddenly,
Out from the flushing centre to the ends
On either hand, their lustrous layers become
Dipt all in crimson streaked with pink and gold;
And then, at last, are edged as with a band
Of crystal fire.

Charles Harpur

Dendrobium gracilicaule
tiger orchid

Some see no beauty in our trees without shade, our flowers without perfume, our birds who cannot fly, and our beasts who have not yet learned to walk on all fours. But the dweller in the wilderness acknowledges the subtle charm of this fantastic land of monstrosities. He becomes familiar with the beauty of loneliness.

Marcus Clarke

Swainsona maccullochiana
ashburton pea

Called the Never-Never, the Maluka loved to say, because they who have lived in it and loved it, Never-Never voluntarily leave it . . . Others – the unfitted – will tell you that it is so called because they who succeed in getting out of it swear they will Never-Never return to it. But we who have lived in it, and loved it, and left it, know that our hearts can Never-Never rest away from it.

Mrs Aeneas (Jeannie) Gunn

Calytrix tetragona
fringe myrtle

THE SICK STOCK RIDER

Let me slumber in the hollow where the wattle
 blossoms wave,
With never stone or rail to fence my bed:
Should the sturdy station children pull the bush
 flowers on my grave,
I may chance to hear them romping overhead.

Adam Lindsay Gordon

Angophora bakeri
narrow-leaved apple

The blossoming of the waratah, the song of the lyrebird, typify the spirit of primitive loveliness in our continent; but the wail of the dingo, the gauntness of our tall trees by silent moonlight, can provide a shiver of terror to a newcomer. Against background of strangeness, of strange beasts and birds and plants, in a human emptiness of three million square miles . . . A new nation, a new human type, is being formed in Australia.

P.R. ('Inky') Stephensen

Grevillea confertifolia
Grampians grevillea

It is not in our cities or townships, it is not in our agricultural or mining areas, that the Australian attains full consciousness of his own nationality; it is in places like this, and as clearly here as at the centre of the continent. To me the monotonous variety of this interminable scrub has a charm of its own; so grave, subdued, self-centred; so alien to the genial appeal of more winsome landscape, or the assertive grandeur of mountain and gorge.

Joseph Furphy ('Tom Collins')
From *Such is Life*

Wahlenbergia stricta
native bluebell

Let us regard the forests as an inheritance given to us by nature, not to be despoiled or devastated, but to be wisely used, reverently honoured and carefully maintained.
Let us regard the forests as a gift, entrusted to any of us only for transient care, to be surrendered to posterity as an unimpaired property, increased in riches and augmented in blessings, to pass as a sacred patrimony from generation to generation.

Ferdinand van Mueller

Sarcochilus falcatus
orange-blossom orchid

WALTZING MATILDA

Oh! there once was a swagman camped in a billabong,
Under the shade of a Coolabah tree;
And he sang as he looked at his old billy boiling,
'Who'll come a-waltzing Matilda with me?'

Who'll come a-waltzing Matilda, my darling,
Who'll come a-waltzing Matilda with me?
Waltzing Matilda and leading a water-bag –
Who'll come a-waltzing Matilda with me?

A.B. (Banjo) Paterson

Eremophila warnesii
emu bush

Beautiful is nature, more or less, almost everywhere; but most grand and impressive where the hand of man has not marred and spoilt it: and this spoiling process, how soon it proceeds with advancing civilisation.

William Strutt

Clematis aristata
old man's beard

THE MAGPIE'S SONG

O, I love to be by Bindi, where the
   fragrant pastures are,
And the Tambo to his bosom takes
   the trembling Evening Star –
Just to hear the magpie's warble in
   the blue-gums on the hill,
When the frail green flower of
   twilight in the sky is lingering still.

Frank S. Williamson

Blandfordia nobilis
Christmas bells

A new heaven and a new earth . . . All creation is new and strange. The trees, surpassing in size the largest English oaks, are of a species we have never seen before. The graceful shrubs, the bright-coloured flower, ay, the very grass itself, are of species unknown in Europe; while flaming lorries and brilliant parroquets fly overhead in the higher fields of air, still lit up in noisy joy, as we may see the gulls do about an English headland.

Henry Kingsley

Gaultheria appressa
waxberry

MOSS ON A WALL

Dim dreams it hath of singing ways,
Of far-off woodland water-heads,
And shining ends of April days
Amongst the yellow runnel-beds.

Deep hidden in delicious floss
It nestles, sister, from the heat –
A gracious growth of tender moss
Whose nights are soft, whose days are sweet.

Swift gleams across its petals run
With winds that hum a pleasant tune,
Serene surprises of the sun,
And whispers from the lips of noon.

Henry Kendall

Anigozanthos flavidus
tall kangaroo paw

Published by the Penguin Group
Penguin Group (Australia)
250 Camberwell Road, Camberwell, Victoria 3124, Australia
(a division of Pearson Australia Group Pty Ltd)
Penguin Group (USA) Inc.
375 Hudson Street, New York, New York 10014, USA
Penguin Group (Canada)
10 Alcorn Avenue, Toronto, Ontario, Canada M4V 3B2
(a division of Pearson Penguin Canada Inc.)
Penguin Books Ltd
80 Strand, London WC2R 0RL, England
Penguin Ireland
25 St Stephen's Green, Dublin 2, Ireland
(a division of Penguin Books Ltd)
Penguin Books India Pvt Ltd
11 Community Centre, Panchsheel Park, New Delhi – 110 017, India
Penguin Group (NZ)
Cnr Airborne and Rosedale Roads, Albany, Auckland, New Zealand
(a division of Pearson New Zealand Ltd)
Penguin Books (South Africa) (Pty) Ltd
24 Sturdee Avenue, Rosebank, Johannesburg 2196, South Africa
Penguin Books Ltd, Registered Offices: 80 Strand, London, WC2R 0RL, England

First published by Penguin Group (Australia), a division of Pearson Australia Group Pty Ltd, 2005

10 9 8 7 6 5 4 3 2 1

Text copyright © Penguin Group (Australia) 2005
Photographic copyright © remains with individual photographers
Photographs p. p. 1, 33, 63 photolibrary.com

The moral right of the author has been asserted.

All rights reserved. Without limiting the rights under copyright reserved above, no part of this publication may
be reproduced, stored in or introduced into a retrieval system, or transmitted, in any form or by any means
(electronic, mechanical, photocopying, recording or otherwise), without the prior written permission of both
the copyright owner and the above publisher of this book.

Cover and text design by Claire Tice © Penguin Group (Australia)
Cover photograph by photolibrary.com
Printed in China by Everbest Printing Co. Ltd

National Library of Australia
Cataloguing-in-Publication data:

Flora Poetica

ISBN 0 1430 0431 X

1. Endemic plants - Australia - Pictorial works.
2. Australia - Pictorial works.

635.95194

www.penguin.com.au